Study Guide for

Alive and Well at the End of the Day

Study Guide for

Alive and Well at the End of the Day

The Supervisor's Guide to Managing Safety in Operations

Second Edition

PAUL D. BALMERT

Published by John Wiley & Sons, Inc., Hoboken, New Jersey.
Published simultaneously in Canada.

For general information on our other products and services or for technical support, please contact our Customer Care Department within the United States at (800) 762-2974, outside the United States at (317) 572-3993 or fax (317) 572-4002.

Wiley also publishes its books in a variety of electronic formats. Some content that appears in print may not be available in electronic formats. For more information about Wiley products, visit our web site at www.wiley.com.

ISBN 978-1-119-90668-1 (paperback)

Cover Design: Wiley
Cover Image: © V. Scott Pignolet / Balmert Consulting

Set in 9.5/12.5 pt STIX Two Text by Straive, Pondicherry, India

SKY10058344_102623

Study Guide

Alive and Well at the End of the Day

"Experience **is** *the best teacher."*

—Anonymous

Alive and Well was designed to serve as a practical guide to safety leadership: its focus is not on explaining leadership *theory* but on teaching leadership *practices*, the "what to do" and "how to do it." In the book, safety leadership practices from starting off the day with a safety meeting to adding up the score at the end of the day are taught in a series of chapters, each one its own, self-contained lesson.

That's how the book teaches the lessons. But life as a leader doesn't come at you like the chapters in a book. So, a more important question for you to ask would be "What's the best way for me to learn, understand, and master the practice of safety leadership?"

Teaching and *learning* may be related, but the processes to accomplish each are entirely different. How do you learn something? More importantly, how do you come to understand something? Most importantly, how do you become really good at something?

Undoubtedly you know the answers.

You might read about something in a book, or hear it explained, but you won't really understand it until you try doing it yourself. That's true for everything from what you learned as a kid, such as tying your shoelaces or throwing a football in a perfect spiral, to what you've learned as a leader in business, like leading a safety meeting or intervening when someone is taking too much risk. Think about what actually happens when you consciously determine to learn how to do something new. As you approach the point of doing something new or different, you essentially teach yourself. That teaching takes place inside your head, normally in the form of a conversation. You first say to yourself, "OK, let's try doing it differently. First, I'm supposed to. . . Next, what I do is . . ."

Then you actually *do* it for the first time. And then you get feedback.

It might work perfectly. You say to yourself, "That was easy." Or it might fail miserably. You think "I'll never try that again." Most of the time, first-time experience falls somewhere in between those extremes. It's different, but far from perfect. You say, "OK, so the next time, I'm going to. . ." Then, you do it again, a little bit differently and a little bit better. You might even get pretty darn good at doing it.

Experience really is the best teacher. We come away from an experience understanding what we know and appreciating what we don't fully understand or can't effectively perform. That is, if—and it's often a really big "if"—we take the time to think about what we have learned. And what we haven't learned. All too often, we're just too busy to take the time to do that.

Study Guide for Alive and Well at the End of the Day: The Supervisor's Guide to Managing Safety in Operations, Second Edition. Paul D. Balmert.

So, learning becomes a hit or miss proposition. We might get lucky, subconsciously getting better at something without really trying. Often, we wind up not learning anything new. That leaves us relying on what we already know. Great if you're satisfied with the present state, but not good if you want to get better. As Einstein put it, "Insanity is doing the same thing over and over again, and expecting different results."

Safety leadership is no different. If you want to get better safety results, you have to lead better. If you want to lead better, you have to first learn first, practice second, and evaluate third. Then do it all over again. That is the continuous improvement process. It is in that process that *understanding* and *mastery* are attained.

Do you want to lead better? If so you need to master the practice of leadership.

In the case of managing safety performance, these practices are described in *Alive and Well at the End of the Day*. Reading and understanding them initiates the learning process. But it will help you get the maximum out of reading the book to realize the real understanding and mastery of the subject takes place at the point where the tools and concepts are put in play: out in the real world and real time of the organization and the people you lead.

This Study Guide is designed to help you master the tools and concepts found in the book as you put them in play. It provides a set of discussion questions you can ask yourself *before* and *after* you've applied the tools and concepts. The questions will help you succeed when you put the tools to use, and better learn from the experience of using them. The questions have been divided into two sections: **Preparation Questions**, designed to ready you to successfully put the tools in play, and **Debrief Questions**, which help you focus on learning from your experience using the tools.

The Study Guide also lays out a structure and process by which you can engage others who have the same interests as you in mastering the tools. Yes, you can follow this Guide on your own, but there's power in numbers. Sharing common experience with others—explaining what you are learning, and listening to their experiences—is guaranteed to enhance your learning.

The Study Guide follows the order of the chapters in the book and provides the kind of structure and process to move from *knowing* something to *understanding* it, to being able to *execute* it effectively. The chapters in the Guide need not be followed in any particular order though. This enables a user to follow a chapter at the time when there is the need. Did you roll out a new safety policy this week? Find the chapter on Communicating Safety Policies. Are you in charge of the department safety meeting this month? Read the chapter on Safety Meetings and see what tools you might use to make the most of the time.

Chapter 1

The Leadership Challenge

No matter where in the world they work, or what kind of business they're in, leaders in operations face the same fundamental set of tough safety challenges. Whether it's getting everyone to follow all the rules all the time, avoiding complacency, getting people to report near-misses, or finding the time to actually *manage* safety performance, these challenges are brutally tough—in large part because they're relentless—and timeless.

The good news is the most successful leaders have figured out how to successfully deal with them, and doing that is largely a matter of practice. . .leadership practice.

Preparation Questions

1. How does the list of "tough safety challenges" on page 4 match up with those you face as a leader in seeing to it that everyone goes home alive and well at the end of the day?

2. What are the factors that make these challenges tough for leaders in operations?

3. What are the specific challenges you are looking for help with?

4. How will you be able to determine how successful you are in dealing with these challenges?

You will find the **Debrief Questions** in the final chapter of the Study Guide.

Study Guide for Alive and Well at the End of the Day: The Supervisor's Guide to Managing Safety in Operations, Second Edition. Paul D. Balmert.

Chapter 2

The Case for Safety

The Case for Safety answers the simple—but critical—question every leader faces: Why is safety your most important goal in operations? It's a question that every leader knows the answer to, but not every leader truly understands that answer. If every leader did, the world of work—and the practice of safety—would look very different.

The answer is found in three simple questions:

What are the most important things in your life?

How would those things be affected were you to suffer a life-changing accident?

Is there anyone working for you whose answers would be different from yours?

Once you understand the answers to these questions—a serious injury could have a devasting effect on your life, and on that point, your followers are no different than you—your mandate becomes clear: Send everyone home safe at the end of every day!

Preparation Questions

1. What are your answers to the first two of those simple questions?

   What are the most important things in your life?

   How would those things be affected were you to suffer a life-changing accident?

2. Who do you know who has suffered a serious accident? What kind of impact did it have on that person and their family?

3. Why do you think so many leaders in operations fail to fully understand *The Case for Safety* until after someone gets seriously injured?

Study Guide for Alive and Well at the End of the Day: The Supervisor's Guide to Managing Safety in Operations, Second Edition. Paul D. Balmert.

4. When you ask those same two simple questions to the members of your crew, how do you think they will answer?

Debrief Questions

1. Who did you pose the questions to?

2. How did their answers compare to what you were expecting to hear?

3. What impact did asking those questions have on their understanding about what makes safety important?

Chapter 3

The Practice of Leadership

What leaders do to lead others in all walks of life boils down to "use words" and "take action." Every follower on the planet knows "actions speak louder than words." That's because actions are things that followers can actually see a leader do—like following the rules and fixing a problem. Still, a substantial portion of leadership activity involves words. Using words—more formally, language—to communicate is a far more complex leadership process: not as important as taking action, but still very important. Words are a leader's way of explaining, eliciting, exciting, and engaging followers.

One very important leadership tool involving words is the "Stump Speech." A Stump Speech is a concise statement of a leader's beliefs, values, advice, and expectations about working safely. It need not be long, but should have enough meaningful content to make it more than just a slogan.

Preparation Questions

In order to develop your Stump Speech for safety, you need to give some thought to its content—what is the message to your followers about working safely? Here are four questions that can help you get your arms around your message:

1. What are the most important reasons the members of your team should work safely?

2. What are the most important things the members of your team should consider or do in order to work safely?

3. What is your best advice about how to work safely?

4. What are the important things you will do as the leader to make sure no one gets hurt on the job?

A Stump Speech has three simple parts:

1. Opening

2. Content: Your key points

3. Closing

Debrief Questions

1. When did you give your Stump Speech?

2. What kind of reaction(s) did you get?

3. What changes—if any—should you consider making? What's missing? Needs clarification? A better explanation?

4. When might be the next good opportunity to put your Stump Speech to effective use?

5. What's the next topic in need of a Stump Speech?

Chapter 4

Moments of High Influence

The concept known as Moments of High Influence is simple—and powerful: in the everyday life of a leader, there are times and places in which the leader is in a natural position of leadership and followers are in a high state of readiness to be influenced. As a leader, when you recognize these moments for what they are—ready-made leadership opportunities—you're bound to raise your game. The secret is to recognize Moments of High Influence in the moment they occur. . .rather than in hindsight. Doing that requires a leader to be able to put theirself in the follower's shoes, seeing the situation from the follower's viewpoint.

Sounds simple. But it is easier said than done!

Preparation Questions

1. Every leader is some other leader's follower. It's easy for a leader to recognize the Moments of High Influence his or her leader has. Why do you think it is often difficult for leaders to recognize their own Moments of High Influence?

2. What safety-related Moments of High Influence have you personally experienced as a follower that have made a big impression on you?

3. What was a recent safety-related Moment of High Influence you had as a leader?

4. What specific safety-related Moments of High Influence do you think you might experience in the next few days?

5. Knowing that each is a Moment of High Influence, what might you do differently than you normally would?

Debrief Questions

1. What was one recent safety-related Moment of High Influence?

2. Recognizing it was a Moment of High Influence, what did you do differently compared to how you would have normally handled the situation?

3. How well did that work?

4. What have you learned from the experience?

5. What else might you do—or do differently—the next time a situation like this comes up?

Chapter 5

Managing By Walking Around

Safety performance always gets better when the supervisor is out on the job seeing what's going on, and just as important, being seen by those doing the work. The challenge is finding the time to actually do that, as supervisors are some of the busiest people on the planet.

That's where Managing By Walking Around fits in: it's the calculated use of the time and presence of the leader. For safety, one of the best ways to perform the calculation is to consider the most likely reasons why people get hurt doing the kind of work your team performs. Once you understand what those most likely reasons are, the best places, times, situations, and people to observe become rather obvious.

Then the game shifts to observation: analyzing the details. What are you seeing that's good. . .and bad?

Preparation Questions

1. What are the top ten reasons why people get hurt doing the kind of work your team performs?

    1.

    2.

    3.

    4.

    5.

    6.

    7.

    8.

    9.

    10.

Study Guide for Alive and Well at the End of the Day: The Supervisor's Guide to Managing Safety in Operations, Second Edition. Paul D. Balmert.

2. To be able to see what real performance looks like for each of these questions, and based on what you know about the work your team performs, how would you structure your Managing By Walking Around efforts (think what, when, where, and who)?

    1.

    2.

    3.

    4.

    5.

    6.

    7.

    8.

    9.

    10.

Debrief Questions

1. Where did you go in your Managing By Walking Around?

2. Why did you choose to go there?

3. What did you see when you were there that was good?

4. What did you see when you were there that was not good?

5. Based on what you saw and learned, where else should you go?

Chapter 6

Following All the Rules All the Time

If you trace every safety rule ever written back to its point of origin, you'll invariably find a tragedy, large or small. In that sense, safety rules are different from all of the other organization's policies and procedures, which are about business. But that difference doesn't make compliance—getting everyone to follow all the rules all the time—easy. Complicating compliance is that there are four separate requirements that need to be met in order for people to comply with the rules.

1. Those expected to follow the rules must first *understand* them.
2. They must be able to *remember* the rules.
3. They must recognize that a rule applies to the situation they find themselves in.
4. They must then choose to *comply* with the rule.

Each requirement poses its own set of unique challenges.

Preparation Questions

1. What specific safety rules do you find your team has the most difficulty following?

2. What are the reasons why compliance with these specific rules is particularly difficult?

3. How would you rate the effectiveness of the training your people receive for these particular rules?

Study Guide for Alive and Well at the End of the Day: The Supervisor's Guide to Managing Safety in Operations, Second Edition. Paul D. Balmert.

4. What actions can you take as a leader to improve compliance with the safety rules?

5. What's your Stump Speech for following all the rules all the time?

6. How can you determine if compliance is improving?

Debrief Questions

1. What specific actions did you take to improve compliance with the safety rules?

2. How well did you execute those actions?

3. What impact have they had on compliance?

4. What else might you consider doing to further improve compliance?

Chapter 7

Recognizing Hazards and Managing Risk

While the terms hazard and risk are sometimes used interchangeably, there is a vitally important difference between them. A *hazard* is a source of danger; *risk* is the probability that the hazard will cause an event. In the case of safety, the consequences of an event can range from nothing more than a minor near-miss to a fatality.

Safety rules are built around known hazards. Unfortunately following the rules doesn't guarantee safety, as there are plenty of ways to get hurt that fall outside the rules. For the hazards most likely to send a follower home hurt or hurting, a supervisor needs to understand what creates a hazard and know where to look for hazards, how to look, and what to look for.

Preparation Questions

1. Most people who get hurt were very knowledgeable about the hazard that caused their injury. Why do you think that is so?

2. Why do you think it is difficult for people to recognize the hazards most likely to send them home hurt—or hurting?

3. In your operation, what do you think are the hazards most likely to wind up sending someone home hurt?

Study Guide for Alive and Well at the End of the Day: The Supervisor's Guide to Managing Safety in Operations, Second Edition. Paul D. Balmert.

4. What are the key hazard recognition and management procedures or processes that you and the members of your team regularly use and rely on to identify and manage hazards?

5. Applying TIE—where to look for hazards—to those policies, how good a job do they do in identifying and managing all the hazards capable of getting someone hurt? What do they miss?

6. What actions can you take as a supervisor to increase the ability of the members of your crew to recognize hazards and properly appreciate their risk—for the tasks they perform?

Debrief Questions

1. What specific actions did you take to better understand how well your hazard and risk identification and management processes are working out on the job site?

2. What did you learn in the process?

3. What does that suggest in terms of actions you need to take to manage the gaps?

Chapter 8

Behavior, Consequences—And Attitude!

As a leader, when you see someone working unsafely—either by not following the rules or taking too much risk—you are called upon to intervene. A successful intervention not only will get the person out of harm's way, it will alter the person's future choice as to their behavior. The intervention strategy summarized in a simple, five-step method, SORRY, represents the approach used by the most successful leaders to accomplish both goals.

Often overlooked in the process of managing behavior is the power of positive feedback to a follower from the leader. With one small modification, the same strategy to correct behavior can be employed to give positive feedback for good behavior. The one change: instead of asking why the person is not working safely, find something specific about the person and the situation that's positive, so as to reinforce safe behavior. Doing that makes the compliment come across as genuine.

A majority of leaders the world over are of the view the best way to get followers to stop working unsafely and start working more safely is to attack what they see as the root of the problem: attitude. It's a great theory, but one that is incredibly difficult to successfully put into practice. If attitude is defined simply as "what someone thinks," it's nearly impossible to know for certain what's going on in a follower's mind. Even if a leader could know, that leaves the bigger problem of changing what a follower thinks. The better strategy is to focus your effort on what you can see, hear, and manage: behavior.

Finally, never lose sight of the fact that consequences do matter to behavior, and the consequences that matter most are found in the Case for Safety: going home alive and well at the end of every day.

Preparation Questions

1. Why do you think changing attitude is such an appealing "target" for so many leaders?

2. What systems and processes (performance appraisals, attitude surveys, for example) does your organization use that focus attention on "employee attitude?"

Study Guide for Alive and Well at the End of the Day: The Supervisor's Guide to Managing Safety in Operations, Second Edition. Paul D. Balmert.

3. How do they compare with those focusing on behavior?

4. If you are struggling with getting compliance on a particular safety policy or procedure, as an alternative to trying to manage attitude, ask yourself this question: What are the *positive* consequences for *not* following the rule?

5. How often are you giving your followers positive feedback in comparison to correcting their behavior?

Debrief Questions

1. When did you last provide safety behavior feedback to someone in your organization? Was it positive or corrective?

2. How closely did you follow the five steps of the intervention strategy?

3. What steps did you find easy to follow? What steps did you find more difficult to follow?

4. The next time you intervene, what will you do better or differently?

Chapter 9

The Power of Questions

Leading by means of asking questions is a great way to gain leadership leverage—increasing your influence as the leader without putting more effort into the process. The way to take advantage of the power found in asking questions is to ask Darn Good Questions.

A Darn Good Question has three elements. The first is purpose: the primary objective of A Darn Good Question is not to get information, but rather to lead and influence a follower with some specific end in mind. The second element goes hand in glove with purpose: a specific audience targeted by the purpose of the question.

The third element is the question itself: A Darn Good Question is based on one of the Key Words: who, what, when, where, how, or why. Starting with any one of these in your question assures an answer other than a simple yes or no. Moreover, each of these words will move your target audience in a specific, different direction as chosen by the leader.

Preparation Questions

1. Why do you think most leaders are more inclined to tell their followers what they think, rather than lead by asking Darn Good Questions?

2. When has a leader you worked for asked you what you now recognize as a Darn Good Question, what impact did that have on you?

3. What are some Darn Good Questions you can ask some of your followers?

Purpose	Audience	Darn Good Question

Debrief Questions

1. What situations did you choose as your opportunity to lead by asking Darn Good Questions?

2. What specific questions did you ask?

3. How well did each specific question work in achieving your purpose?

4. Based on your experience, what have you learned about the process of asking Darn Good Questions?

Chapter 10

Making Change Happen

Every time a supervisor stands up in front of the crew, announcing that, "We have a new safety policy," it's a Moment of High Influence. But communicating change will almost always generate some amount of resistance simply because a follower doesn't have any choice but to go along with the change. Followers don't resist change. . .they resist *being* changed.

Assuming the change affects your followers, as the supervisor, you are responsible for the most important step in the change process: execution, making the change happen. Whenever a policy is changed, the supervisor's goal is to have every team member understand what the change is, and be committed to making the change.

Understanding that, there are two simple rules to follow that will make the change process go better. The first is to explain why the change is being made. Explaining why reduces resistance as people are far more likely to go along with a change when they know the reason. In the case of safety policies and procedures, the reason for the change almost always has something to do with something bad that happened somewhere else. The second rule is to focus on making the change happen, by asking, "What do we need to do to make the new policy work?"

Preparation Questions

The perfect time to apply this chapter is at the point where you will be communicating a new or revised safety policy or procedure. When you're called upon to do that, consider the following questions before you communicate the change.

1. What caused the policy to be written or changed?

2. What issues or questions can you expect to be raised by those who are impacted by the change?

Study Guide for Alive and Well at the End of the Day: The Supervisor's Guide to Managing Safety in Operations, Second Edition. Paul D. Balmert.

3. How will you begin to communicate the change?

4. How will you close out the session and assure understanding and commitment to the change?

Debrief Questions

1. How well did you follow your outline for communicating the change?

2. What steps in the process went well?

3. What problems did you encounter in explaining the change, getting buy-in, or following your communication plan?

4. What have you learned from the process that will help you the next time you communicate a change?

Chapter 11

Understanding What Went Wrong

When the consequences of a problem are big, a formal investigation is launched. Fortunately, those events are relatively rare, leaving looking into the everyday, ordinary looking problems—often with little to no consequence—to the discretion of a supervisor. But often the only real difference between a major disaster and a non-event is found in its consequences. Problems with little or no consequence can be the best opportunities to learn about what's really going on in your organization and deal with situations before they become big problems.

At the core of every good investigation are answers to the Fundamental Questions: who, what, when, where, how, and why. The answers to all but the last question—why—are found in the facts. The simple approach is to find the facts first, and save the answers to the question why—conclusion, judgment, opinion—until after you've established the facts.

Understanding what went wrong is only half the battle; coming up with a solution that actually solves the problem is every bit as important. A good solution is one that meets two tests: 1) effectiveness, how well the solution fixes the problem; and 2) efficiency, how much time, effort, and resources are required to implement the solution.

Finally, never forget that every investigation meets the definition of a Moment of High Influence.

Preparation Questions

1. How often do you get reports of "minor safety problems" that don't reach the threshold of requiring a formal investigation? What kind of problems do you typically hear about?

2. Of those problems, which ones do you think represent the best opportunities to learn more about what's really going on in your organization, or to fix problems before they produce significant consequences?

Study Guide for Alive and Well at the End of the Day: The Supervisor's Guide to Managing Safety in Operations, Second Edition. Paul D. Balmert.

3. What types can you single out as the best opportunities to better understand what went wrong?

4. What's your plan to take full advantage of one of these potential opportunities?

Debrief Questions

1. What problem did you choose to investigate?

2. How well did you follow the process of asking the Fundamental Questions?

3. What did you learn about understanding what went wrong and solving problems?

4. What will you do different and better the next time?

Chapter 12

Managing Accountability

Every leader knows "holding people accountable" is an important part of managing safety performance. Successfully managing accountability begins by understanding what the term really means. Then it requires understanding the steps in the process to "hold someone accountable."

Managing accountability is undertaken on the basis of a leader's understanding of the cause as to what went wrong, and what a follower did to contribute to the problem. It's a discussion with a follower lead by a leader taking place in the aftermath of a failure. It's a two-way conversation intended to change future behavior for the better.

As to the process of holding a follower accountable, it is about asking the questions that need to be asked to establish ownership for what was done wrong, the consequences—actual and potential—for what went wrong, reaffirming expectations as to the follower's duties and responsibilities, and setting potential future consequences should behavior not change for the better.

The Five Ss provide the structure for your discussion to manage accountability. The first three S's—situation, significance, and specifics—set the stage for the core of the process: asking the questions that need to be asked. The natural place to start is by asking the person for "their side of the story," and then, of course, digging deeper by asking Darn Good Questions. Finally, holding someone accountable must include identifying the steps to prevent the problem from happening again. That's the final step in the process.

Preparation Questions

1. "Managing accountability" is often confused with the administration of corrective action or discipline. They are related, so it's important to understand the latter as well. In your organization, what are the *potential* consequences for not following the safety rules or taking too much risk?

2. Thinking back over your experience, when have you attempted to "hold someone accountable," as opposed to "administering consequences" in common forms such as counseling, oral or written warnings and even time off?

Study Guide for Alive and Well at the End of the Day: The Supervisor's Guide to Managing Safety in Operations, Second Edition. Paul D. Balmert.

3. How well did that approach work? How closely did it follow each of the Five Ss?

4. Looking forward, what specific situations do you foresee which lend themselves to holding a follower accountable?

5. If you follow the Five Ss, what will you say, or ask, at each step in the process?

Debrief Questions

1. What was the specific situation for which you followed the Five Ss to hold a follower of yours accountable?

2. How well did you follow each of the steps in the process?

3. As you followed the process, what steps proved easy? What steps were more challenging?

4. What were the important questions you asked in the fourth step of the process? What questions worked well? What ones could have worked better?

5. The next time you hold someone accountable, what can you do even better?

Chapter 13

Managing Safety Suggestions

When a supervisor gets a safety suggestion, it's a Moment of High Influence. . .for the employee who makes the suggestion. As for the leader, it's another opportunity to lead. It also means dealing with a problem, because a safety suggestion represents the combination of a problem and a solution. Not that the supervisor necessarily gets both parts.

The first consideration in dealing with any safety suggestion is that of priority or urgency: how quickly the suggestion must be dealt with. That determination should be based on the nature and severity of the problem. Next, you always want to say "thank you" (not "good idea") to the person who has submitted the suggestion: that's how you recognize the Moment of High Influence.

Finally, as to what to do about the suggestion, there are always plenty of options, such as saying, "Let's do it," counter-proposing an alternative solution to the problem, looking at the problem yourself—otherwise known as Managing By Walking Around—delegating the problem to someone else, turning in a formal safety suggestion, and letting the employee champion the idea. If the solution finds its way into practice, be sure to bring it to the attention of the rest of the crew, giving proper credit to the person who made the suggestion. That's a great way to encourage others to turn in suggestions.

Preparation Questions

1. How often do the members of your crew turn in safety suggestions?

2. How is a safety suggestion normally handled in your operation?

3. How well does that process work?

4. What kind of safety problems would you like to see brought up by suggestion(s) from your crew?

Study Guide for Alive and Well at the End of the Day: The Supervisor's Guide to Managing Safety in Operations, Second Edition. Paul D. Balmert.

5. The next time you receive a safety suggestion, what will be your plan to respond?

Debrief Questions

1. What was one specific safety suggestion you received?

2. What was the problem? The proposed solution—if you received one?

3. What specifically did you do—your words and actions—when responding to the suggestion?

4. What went well?

5. What might you do differently the next time you receive a safety suggestion?

Chapter 14

Safety Meetings Worth Having

A safety meeting—whether it's the morning tool box safety meeting, the monthly department safety meeting, or the annual "all hands" meeting—is in theory a Moment of High Influence. But, if it's not a good safety meeting, what kind of influence will it have out on the job? Worse, what kind of message does a leader send to followers by leading a bad safety meeting?

One sure way to run an effective safety meeting is to get your audience to do the talking for you. Of course, you want them talking about what *you* want them to—not what *they* would like to—and, in the process, actually accomplish something important.

An Ask, Don't Tell safety meeting is a simple and effective way to do exactly that.

Like asking a Darn Good Question, this process begins with the leader being very clear and specific as to exactly what they want to accomplish by talking about the subject. The next step is to keep the story short. Details are seldom necessary to your purpose, meaning it should take but a few moments to set up the subject with a headline and summary. Before asking Darn Good Questions, make the connection by explaining why your followers should pay attention to the topic and participate in the discussion. Then, two or three Darn Good Questions are normally sufficient to produce 15 minutes of valuable discussion on the subject.

Preparation Questions

Pick out a topic for an Ask Don't Tell Safety Meeting, and then complete the outline.

Purpose: *My goal in getting my team to talk about this subject is so that...*

Headline:

Summary:

Study Guide for Alive and Well at the End of the Day: The Supervisor's Guide to Managing Safety in Operations, Second Edition. Paul D. Balmert.

Connection: "*The reason why it is important for us to talk about this topic today is because....*

Darn Good Question 1:

Darn Good Question 2:

Darn Good Question 3:

Debrief Questions

After you've run a safety meeting following the Ask Don't Tell method, answer these questions:

1. What topic did you choose? What was your purpose?

2. In the meeting, how long did each of the sections (headline, summary, connection and questions) take?

3. How closely did the meeting follow your Ask Don't Tell plan?

4. What worked well?

5. What Darn Good Questions did you ask? What questions worked well? What questions did not work as well as you wanted?

6. The next time you use Ask, Don't Tell, what will you do differently?

Chapter 15

Creating the Culture You Want

Every operation has its safety culture: "What most people do most of the time." If you want to know what kind of safety culture your operation has, all you have to do is look and listen—not at what any one person does, but at what "most people do." It isn't necessary to look at everything. Pick out a few key dimensions, like people stopping at the stop signs, carefully executing job safety analyses, and reporting near-misses. . .and then observe. What you observe is your safety culture. If what you see is not a "culture of safety," you then need to be in the culture *change* business.

Left unattended, a culture will slowly change but in a direction of its choosing, not yours. You want *transformation,* which implies both speed and a change in a direction of your choosing. For that you need a roadmap.

The road to transforming culture starts with clarity about the change you want. Bear in mind, changing culture means changing behavior, the way work gets done. Next, look for leverage: key people with outsize influence and small things that can play large. Then get into the selling mode: what's in it for your followers to buy into the changes you want. Finally, look for the Moments of High Influence, when your followers are paying attention to you. Seize the Moment!

Preparation Questions

1. In terms of specific and observable behavior, how would you characterize your operation's current safety culture?

2. If you don't have a good sense of what that culture is, where and when and how should you look to gain this insight?

3. What would you like your safety culture to be, described in specific terms of behavior?

Study Guide for Alive and Well at the End of the Day: The Supervisor's Guide to Managing Safety in Operations, Second Edition. Paul D. Balmert.

Debrief Questions

1. Based on your data and first-hand observation, what are the key targets for change that you see as critical to creating the safety culture you want?

2. What are the key leverage points—people, places, events—you can take advantage of to speed up the change process and increase the likelihood that you will create the safety culture *you* want?

3. How will you measure your progress, and your success in creating the culture you want?

Chapter 16

Investing In Training

Knowledge is the single most important factor in going home alive and well, putting training front and center in managing safety. Sixty years ago, McGehee and Thayer summed up the potential return on investment in training, "The effectiveness of achieving an organization's goals will depend, in a significant way, on the nature and efficiency of the training employees receive for their assignments." What's changed in the years since? Only the breadth and depth of what people need to know to be able to meet those goals, starting with knowing what they need to know to be able to work safely. Every change creates the need for new knowledge. The need for understanding has never been greater.

Knowledge is no guarantee that anyone will work safely. But it falls to the leader to ensure that every follower knows what they need to know to be able to work safely. If they don't, it's not the training department's problem. . .it's your problem.

Put what your people know to the test. If you're not satisfied that they understand what they need to know, it's time to train. And, if training is what's needed, keep the Three T's in mind: timing, technique, and teacher. Properly combine all three, the result can be memorable training; the kind of training that creates understanding, and will help you achieve your goal to send every follower home alive and well, every day.

Analysis Questions

1. What are the key things your followers need to know to be able to do their jobs safely?

2. How well does each follower measure up against what they need to know?

Study Guide for Alive and Well at the End of the Day: The Supervisor's Guide to Managing Safety in Operations, Second Edition. Paul D. Balmert.

3. How well does each existing training program perform in delivering the knowledge your followers require?

4. Since it's not possible to take on every training challenge at the same time, what are the highest priority topics and courses in need of improvement in timing, technique and teaching?

5. What's the best path forward to improve training in these key need areas?

Chapter 17

Measuring Safety Performance

To manage safety performance effectively, every leader needs accurate information to be able to judge and evaluate performance, determine the future direction of performance, and reveal to the leader what's really going on. How the leader uses that information determines its function. A useful leading indicator will reliably predict a future change in the trend of performance, explaining both its desirability *and* difficulty. When followers know certain information will be used by their leaders to judge performance, they will be naturally inclined to manage the information to produce the result the leader wants. Every leader must understand that type of information cannot be relied upon to reveal reality; other methods must be found to meet that vital need.

As to how to go about improving the measurement process to produce better information on which to manage, the scope of that effort depends on the level of the leader: a front line leader needs only information concerning their crew, whereas the executive's scope is the larger organization. In either case, it's best to start with what already exists; scouring the organization for information and metrics will likely produce a long list of information routinely collected and metrics that have been created. For an executive, all that information can be evaluated and understood by applying the technique of the Balanced Scorecard. For the front line leader, the practical approach is creating Early Warning Indictors.

Assessing Current Practice

1. What are the current safety performance metrics and information your organization collects?

2. How is each principally used by leaders: judge, predict, or reveal?

3. How do they collectively fit into the Balanced Scorecard?

Study Guide for Alive and Well at the End of the Day: The Supervisor's Guide to Managing Safety in Operations, Second Edition. Paul D. Balmert.

4. Which ones are good candidates to serve as Early Warning Indicators?

5. What *don't* you know about the safety performance of the followers you lead that you think you should?

6. Given your assessment of the current state of safety performance measurement, what are the opportunities to improve how safety performance is measured?

Chapter 18

Managing Safety Dilemmas

A dilemma is a situation where two conditions exist and are in opposition to each other. Managing safety performance is made even more difficult by the fact that leaders routinely find themselves on the horns of six significant dilemmas:

- The Accountability Dilemma: The gap between what a leader is accountable for, and what a leader actually can control or influence.
- The Risk Dilemma: Operating with zero risk is impossible. Yet, when an injury occurs, the consequences are unacceptable.
- The Investigation Dilemma: Finding out *what* went wrong invariably leads to determining *who* did something wrong. But nobody wants to be found out to be that person.
- The System Dilemma: Individual human performance is a function of system performance. But, every individual is free to choose how he or she behaves.
- The Middle Dilemma: The middle level of management in the hierarchy plays the vital role of linking those at the top with those that do the work. Yet, leaders in the middle of any organization feel powerless, caught in the crossfire between those two levels.
- The Leader Dilemma: In producing great results, many highly effective leaders are quiet and self-effacing. Yet, that is likely to leave the best leaders under-recognized and under-appreciated in the organizations they work for.

Managing these safety dilemmas is never easy. There's no getting around the fact that a dilemma pulls a leader in opposing directions, and it's all but impossible to eliminate either of their "horns." Dilemmas begin to become manageable when a leader first understands where the competing forces come from, and then develops a strategy for dealing with the dilemma.

Preparation Questions

1. In the past, when have you faced (or seen another leader face) these safety dilemmas:
    - The Accountability Dilemma
    - The Risk Dilemma
    - The Investigation Dilemma
    - The System Dilemma

Study Guide for Alive and Well at the End of the Day: The Supervisor's Guide to Managing Safety in Operations, Second Edition. Paul D. Balmert.

- The Middle Dilemma
- The Leader Dilemma

2. What made dealing with the dilemma you faced difficult for you as a leader?

3. What dilemma are you most likely to face next?

4. The next time you face the dilemma, what will be your strategy to deal with it?

Debrief Questions

1. What specific dilemma did you face?

2. What were the competing forces that created the dilemma—for you?

3. What did you do to deal with the dilemma?

4. What worked well? What didn't go as well as you would have liked?

5. What have you learned that you can put to good use the next time this dilemma comes up?

Chapter 19

Leading From The Middle

The Leadership Model suggests that leadership takes place in a downward direction in the organization: to the followers who report to the leader. But there are situations where the leader's target of influence are peers, customers, and sometimes the leaders they report to. That's known as Upward Leading, where the principles of leadership are unchanged, but the practical realities make the practice challenging.

Leaders often feel powerless, but that sense reflects a lack of understanding of Organization Power, the degree of control or influence a leader has in a situation. Applying Organization Power to a list of the Critical Safety Factors that have caused many high profile safety events, by virtue of stop work authority—control—the front line leader is often the most powerful member of management in the hierarchy. A front line leader has two other very significant sources of influence: information as to what's really going on, and credibility with followers.

The proper understanding and application of Organization Power won't make a leader's life easier, but is will make safety performance better.

Preparation Questions

1. Why do you think many front line leaders fail to appreciate the Organization Power they actually possess?

2. One particularly important opportunity to "lead from the middle" is to provide upper management with a true picture of organization reality as it relates to safety. In your organization, where do you see a potential disconnect between what your leaders think is going on, and what is really going on?

Study Guide for Alive and Well at the End of the Day: The Supervisor's Guide to Managing Safety in Operations, Second Edition. Paul D. Balmert.

3. What can you do to “upward lead” your leaders by providing a more realistic picture of organization performance?

4. What’s your strategy to “upward lead”? Who? When? How?

Debrief Questions

1. What was the opportunity you had to “upward lead?”

2. What did you do? What tools did you use?

3. How well did it work out? What was the reaction by your leader?

4. Based on your experience, what will you do next time? What will you do differently?

Chapter 20

Mistakes Managers Make

When it comes to mistakes, leaders are no different than their followers: they've made their share of mistakes. These mistakes are often made by well-intentioned leaders, and every mistake represents an opportunity to learn. But those lessons come at a cost, and when they involve safety, that cost can be harm to followers. The smarter strategy is to learn from the mistakes of other leaders. But that requires the time to find out about them, to study them and the willingness to admit, "I could make that same mistake."

Here are six Big Mistakes managers make in managing safety performance:

1. Failing To Prepare
2. Driving Out All Fear
3. Focusing On The Short Run
4. Trying To Buy A Game
5. Confusing Perception With Reality
6. Communicating Without A Common Vocabulary

Discussion Questions

1. As a leader, what big mistakes have you made managing safety?

2. What mistakes have you seen other leaders make?

Study Guide for Alive and Well at the End of the Day: The Supervisor's Guide to Managing Safety in Operations, Second Edition. Paul D. Balmert.

3. Why do you think you—or they—made the mistakes?

4. In your role as a leader, where do you think you might be inclined to fall victim to the same thinking that produced the mistake?

Chapter 21

Driving Execution

Execution is defined as "the doing part of every work process." Sending every follower home, alive and well at the end of the day boils down to execution, defined as "the doing part of any work process." Driving safety execution is the job of every leader, from front-line supervisor to CEO. The Four Common—and Wrong—Assumptions help a leader understand why execution is normally far from flawless. The Four Absolute Truths About Execution begin to provide answers to what every leader needs to do to manage it.

Unless safety execution is flawless, there is bound to be an Execution Gap. How well a leader understands that gap is found in Performance Visibility. Taken together, the leadership practices described throughout *Alive and Well* are all about execution. They provide the "what to do" and "how to do it" to achieve the level of execution necessary to send everyone home safe. That is always the most important duty of every leader.

Preparation Questions

1. How would you rate the current level of "safety execution" in your operation? How big is the Execution Gap? Where does it exist?

2. If you aren't certain as to what your operation's "safety execution" really is, how can you objectively find out?

3. For the areas and activities for which safety execution is not as good as you want it to be, what are the most important things to focus on improving?

Study Guide for Alive and Well at the End of the Day: The Supervisor's Guide to Managing Safety in Operations, Second Edition. Paul D. Balmert.

4. As the leader, what is your plan to improve execution in those key areas and activities?

5. How will you measure improvement in execution? What Early Warning Indicators can you use to tell you how much progress is being made?

Debrief Questions

1. What are the specific areas and activities you selected to improve execution in?

2. What techniques from *Alive and Well* did you utilize?

3. What specific actions did you take as a leader?

4. What worked well?

5. What could have gone better?

6. What improvements in execution have you seen?

7. What's next to further improve execution, or keep it at the level of excellence you have achieved?

Chapter 22

Making a Difference

Taken together, the first 21 chapters in the book provide practical answers to the four questions every leader faces:

1. What is leadership?
2. When do I lead?
3. How do I lead?
4. Why must safety always come first?

Armed with the answers, it falls to the leader to make a decision: what am I going to do about all those tough safety challenges I face every day as a leader?

When it comes to sending people home safe, leadership is the difference that makes the difference. There's no getting around that fundamental truth. Getting "common men to do uncommon things"—as Peter Drucker put it—is the stuff of great leadership.

Final Questions

1. What have you learned about managing safety performance from reading and discussing *Alive and Well at the End of the Day?*

2. What tools and concepts have you put into play that have helped lead the members of your crew to work more safely. What have you learned from applying these tools?

3. What will you do next to make a difference in the most important goal every business has. . .sending everyone home, alive and well, every day?

Study Guide for Alive and Well at the End of the Day: The Supervisor's Guide to Managing Safety in Operations, Second Edition. Paul D. Balmert.